Learn About The
BALD EAGLE

A Curious Kids' Picture Book

By Laura Linwood

Greenlawn Publishing

Dedicated with loving kindness
to my grandchildren, Calvin and Julia.

May your curiosity always grow and
may learning always feel like an adventure.

CONTENTS

INTRODUCTION

Real animals. Real photographs. Real facts. Every book in this series takes a child deep into the world of one amazing animal – how it looks, how it moves, how it feeds, and how it lives.

We think the real world is more astonishing than anything we could make up. So every page pairs a stunning, true-to-life photograph with a surprising fact – the kind that makes a young reader stop and look closer. It's made to be read out loud, wondered over together, and opened again and again.

At the back, eight activities – mazes, word searches, and puzzles – turn what you've discovered into play. And there's more waiting online: find bonus printable activities, photos, videos, and more for every animal at curiouskidslearning.com.

LOOK CLOSELY

The following photographs deserve your close attention.

Each one is paired with surprising facts. Look carefully – the more you look and read, the more you discover.

For parents: Read each fact aloud as you explore the photos together. Ask your child what they notice first.

Photo credits and bonus content can be found at curiouskidslearning.com. Scan QR code on the last page.

TOP OF THE WORLD

FROM HIGH UP HERE, THE BALD EAGLE CAN SEE FAR BELOW TO RIVERS, OCEANS, AND FIELDS – SCANNING FOR PREY FOR ITSELF OR ITS YOUNG.

Look at where this eagle is. Snow on the mountains. Valley far below. Wind up here that would knock you sideways. And this eagle is completely at home.

Bald eagles live near water – rivers, lakes, and coastlines – but they perch high above everything, watching. From up here nothing moves without the eagle knowing. Nothing flies past without being seen. This is an animal that owns its world entirely.

MIGHTY WINGS

THOSE WINGS STRETCH MORE THAN SEVEN FEET TIP TO TIP – TALLER THAN A STANDARD DOORWAY.

Next time you walk through a doorway, look up at it. That's how wide a bald eagle's wings stretch in full flight. More than seven feet.

And here's something that stops most people cold – those feathers weigh more than the eagle's entire skeleton. Every bone in its body hollowed out and lightweight. Every feather dense and precise. About 7,000 of them, each one shaped for a specific job.

The eagle finds a current of warm air rising from the earth, tilts its wings slightly, and rises without a single flap. Miles traveled. Barely any effort at all.

EYES ON TARGET

A BALD EAGLE CAN SEE A FISH SWIMMING BENEATH THE SURFACE OF A LAKE FROM ALMOST A MILE AWAY – AND IT CAN SEE COLORS AND LIGHT HUMAN EYES CAN'T DETECT AT ALL.

Look at those eyes. Locked forward. Unblinking. Already decided. Eight times sharper than yours.

A bald eagle sees in ultraviolet light – a whole range of color invisible to humans. It uses this to track the trails small animals leave on the ground far below. A rabbit in a field from two miles away is perfectly clear. A fish beneath the surface of a lake from almost a mile away – already located.

By the time an eagle looks like this, the target has been chosen. The rest is almost automatic.

TALON POWER

THOSE TALONS GRIP TEN TIMES STRONGER THAN THE STRONGEST HUMAN HAND – AND THEY LOCK THE INSTANT THEY MAKE CONTACT.

Those aren't just claws. They are the most powerful tools this eagle owns. Curved. Sharp. Built to close on impact and stay closed. Ten times stronger than a human grip.

The eagle leads with its feet – talons forward, coming in fast, not slowing down. This is what the hunt looks like from the other side. Whatever those talons close around stays closed around. There is no escape from this moment.

THE STRIKE

A BALD EAGLE DIVES AT UP TO 100 MILES PER HOUR AND HITS THE WATER FEET FIRST – AND AT THE LAST INSTANT A SEE-THROUGH EYELID CLOSES OVER EACH EYE LIKE GOGGLES.

This is it. Feet forward. Water exploding. Fish grabbed. 100 miles per hour and it hits with pinpoint accuracy. Those built-in goggles protect the eyes on impact without blocking the view.

The eagle sees the fish clearly right up until the moment of contact. The whole strike – from dive to catch – takes less than a second.

Almost every single time, the eagle comes away with exactly what it came for.

NEVER LETTING GO

AN EAGLE'S TALONS LOCK AUTOMATICALLY ON IMPACT – IT COULDN'T OPEN THEM EVEN IF IT WANTED TO.

The fish never had a chance.

Look at those talons – locked around it, dripping, rising out of the green water. The grip engaged the instant contact was made. No thought. No effort. Just locked.

Sometimes the fish is so heavy the eagle can't lift it into the air. So it rows itself to shore instead – using its wings like oars, pulling across the surface of the water. It will not let go. It cannot let go. That fish is coming with it no matter what.

NOT THERE YET

THAT BROWN BIRD ISN'T A DIFFERENT SPECIES – IT'S A BALD EAGLE, JUST YOUNGER. AND IT HATCHED WITH A TINY SPIKE ON ITS BEAK TO BREAK OUT OF ITS SHELL.

See the brown bird next to the white headed eagle? Same species. Same family.

Young bald eagles are completely brown for their first few years. Slowly, patch by patch, the white feathers begin to appear. By around age five the transformation is complete. That famous white head finally arrives.

And before any of that – before the brown feathers, before the first flight – there was a tiny spike on the end of a beak. Just strong enough to crack open a shell. Just long enough to be needed once. Then gone.

INTO THE SKY

WHEN AN EAGLE LOSES A FEATHER ON ONE WING, ITS BODY AUTOMATICALLY DROPS A MATCHING FEATHER FROM THE OTHER WING – TO KEEP IT PERFECTLY BALANCED IN FLIGHT.

Watch that launch. One push off the branch. Wings out. Air caught. And from this moment on, almost no flapping at all.

The eagle finds the invisible rivers of warm air rising from the earth and lets them do the work. Up and up in wide lazy circles, scanning everything below.

Every one of those 7,000 feathers doing exactly its job. And if one falls out – the other wing automatically drops one too. Perfectly matched. Perfectly balanced. Every single flight.

FOR LIFE

BALD EAGLES MATE FOR LIFE – SAME PARTNER, SAME NEST, EVERY SINGLE YEAR – AND IN WINTER HUNDREDS OF THEM GATHER TOGETHER NEAR OPEN WATER.

One is staying. One is leaving. But they will be back together soon. Bald eagles choose one partner and keep that partner for life.

Year after year they return to the same nest – adding sticks, making it stronger, making it more home than it was the year before.

And when winter comes and the lakes start to freeze, eagles gather. Hundreds of them, near the open water where fish are still reachable.

Same birds. Same places. Same loyalties. Every year.

THE BIGGEST NEST

THAT NEST CAN WEIGH OVER A TON — HEAVIER THAN MOST CARS — AND SOME HAVE BEEN USED BY THE SAME EAGLES FOR DECADES, GROWING BIGGER EVERY SINGLE YEAR.

Look at the size of that nest. Sticks and branches piled deep — some as thick as your arm.

A bald eagle's nest is the largest built by any bird in North America. Each year the eagles come back and add more. More sticks. More weight. More home.

Over decades some nests grow so large and so heavy that the tree beneath them eventually gives way. And right now this eagle is coming back to it. Wings out. Feet down. Almost home.

GONE AND BACK

IN THE 1960S BALD EAGLES NEARLY VANISHED FROM THE EARTH – A PESTICIDE CALLED DDT MADE THEIR EGGSHELLS SO THIN THEY CRUMBLED BEFORE EAGLETS COULD HATCH.

Still. Watching. Unhurried.

This eagle has no idea how close its kind came to disappearing forever. By the 1960s there were almost none left. DDT was destroying the Bald Eagle population. People noticed. Laws changed. The pesticide was banned. Slowly, one nest at a time, the eagles came back.

Today more than 300,000 bald eagles live across North America.

This one is perched in an evergreen, watching the world go by, completely unaware that its very existence is a miracle.

EYE TO EYE

THOSE EYES ARE EIGHT TIMES SHARPER THAN YOURS – RIGHT NOW IT SEES YOU MORE CLEARLY THAN YOU SEE IT.

Look right into those eyes. Go ahead. Don't look away. A bald eagle's eyes are nearly as large as a human's – packed into a head much smaller than yours. Eight times sharper than anything you can see.

Right now, in this moment, it is looking straight back at you. It sees every detail of your face. It sees the room behind you. It sees everything. And it has seen you long before you saw it.

FUN ACTIVITIES

Eight activity pages give you two ways to play with what you've learned — one built for younger explorers, one built for older adventurers.

Start with the first. If it's too easy, try the second. If the second stumps you, come back to it later.

For parents: Each activity has an easy version for ages 4–6 and a challenge version for ages 7–10. Both versions are in here on purpose.

Bald Eagle Maze I

Find your nest!

Bald Eagle Maze II

Catch your dinner!

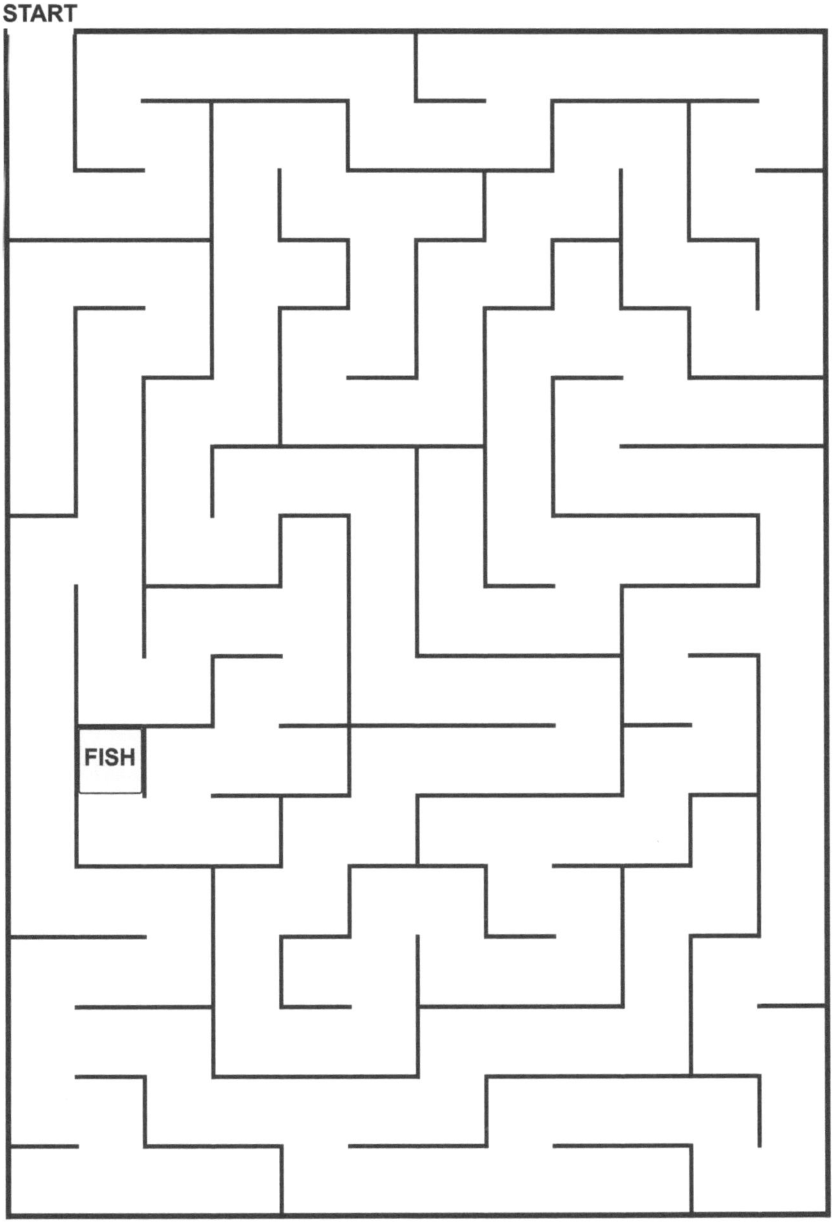

Bald Eagle Word Search I

Find and circle each word — across and down!

S	U	B	U	B	J	Q	S	N	F
K	H	E	L	T	R	G	E	C	S
W	B	A	W	E	G	G	S	L	K
C	B	K	G	Y	W	V	P	A	K
D	I	V	E	Y	D	Y	E	W	J
E	V	C	M	O	N	M	R	W	S
P	C	H	I	C	K	W	C	I	J
S	O	A	R	T	P	Q	H	N	R
I	L	F	Q	E	Y	E	S	G	H
U	N	E	S	T	A	A	U	D	K

Word List

BEAK	CHICK	CLAW	DIVE
EGGS	EYES	NEST	PERCH
SOAR	WING		

Bald Eagle Word Search II

Find and circle each word — across, down, and diagonal!

F	U	P	R	H	X	P	V	A	N
A	R	T	I	A	E	R	G	E	A
L	A	E	V	T	A	E	S	R	S
C	P	R	E	C	G	D	C	I	H
O	T	R	R	H	L	A	A	A	O
N	O	I	M	S	E	T	T	L	R
Z	R	T	W	D	T	O	C	Q	E
K	F	O	X	R	G	R	H	J	L
P	O	R	Q	S	H	A	R	P	I
P	K	Y	E	Y	R	I	E	U	N
H	A	T	C	H	L	I	N	G	E
A	C	O	U	R	T	S	H	I	P
P	L	U	N	G	E	C	V	H	B
R	U	M	O	L	T	I	N	G	O
W	C	A	R	R	I	O	N	R	L
L	C	L	U	T	C	H	B	G	D

Word List

AERIAL	BOLD	CARRION	CATCH
CLUTCH	COURTSHIP	EAGLET	EYRIE
FALCON	HATCH	HATCHLING	MOLTING
PLUNGE	PREDATOR	RAPTOR	RIVER
SHARP	SHORELINE	SWOOP	TERRITORY

Bald Eagle Scramble I

Unscramble the letters to make a word.

---

1. **WNGI**

Answer: _ _ _ _

2. **YESE**

Answer: _ _ _ _

3. **RHECP**

Answer: _ _ _ _ _

4. **YEPR**

Answer: _ _ _ _

5. **ICCKH**

Answer: _ _ _ _ _

6. **EAKB**

Answer: _ _ _ _

Word Bank

BEAK	CHICK	EYES
PERCH	PREY	WING

Bald Eagle Scramble II

Unscramble the letters to make a word.

---

1. **LAERAI**

Answer: _ _ _ _ _ _

2. **OPRART**

Answer: _ _ _ _ _ _

3. **SBLMYO**

Answer: _ _ _ _ _ _

4. **LOANT**

Answer: _ _ _ _ _

5. **RCPHE**

Answer: _ _ _ _ _

6. **DEPES**

Answer: _ _ _ _ _

7. **KEREC**

Answer: _ _ _ _ _

8. **NUEGPL**

Answer: _ _ _ _ _ _

Word Bank

AERIAL	CREEK	PERCH
PLUNGE	RAPTOR	SPEED
SYMBOL	TALON	

Bald Eagle Fill-In I

Fill in the blank to complete each fact.

---

1. The bald eagle is known for its bright white head, bold yellow ________, and dark brown wings.

   Answer here:

2. Bald eagles lock their ________ around slippery fish , they rarely drop their catch.

   Answer here:

3. A bald eagle's ________ are nearly as large as a human's , and up to eight times sharper.

   Answer here:

4. Bald eagles mate for life and return to the same ________ each year, adding new sticks.

   Answer here:

5. A bald eagle can spot a ________ swimming in a lake from almost a mile away.

   Answer here:

6. Bald eagles ________ feet-first and snatch fish right off the surface of the water.

   Answer here:

7. A bald eagle's wingspan can stretch over 7 feet wide , wider than most ________ are tall.

   Answer here:

8. The bald eagle is the ________ bird of the United States , a symbol of strength and freedom.

   Answer here:

Word Bank

BEAK	DIVE	EYES
FISH	GROWN-UPS	NATIONAL
NEST	TALONS	

Bald Eagle Fill-In II

Fill in the blank to complete each fact.

---

1. The bald eagle is the national bird of the United States , a ________ of strength and freedom.

   Answer here:

2. The bald eagle is known for its bright white head, ________ yellow beak, and dark brown wings.

   Answer here:

3. Bald eagles lock their talons around slippery fish , they rarely drop their ________.

   Answer here:

4. A bald eagle's nest, called an ________, is the biggest of any bird in North America.

   Answer here:

5. Eagle chicks are called eaglets. They ________ covered in soft gray fluff, not feathers.

   Answer here:

6. A bald eagle's ________ can stretch over 7 feet wide , wider than most grown-ups are tall.

   Answer here:

7. Bald eagles live near water , rivers, lakes, and ________ are their favorite hunting grounds.

   Answer here:

8. Bald eagles were nearly extinct. Today, thanks to ________, over 300,000 live in North America.

   Answer here:

Word Bank

BOLD	CATCH	COASTLINES
EYRIE	HATCH	PROTECTION
SYMBOL	WINGSPAN	

MORE TO EXPLORE

Your child's wonder and curiosity about the Bald Eagle doesn't have to end here.

Scan the QR code to discover additional resources that spark new adventures.

- Want even more Bald Eagle fun? Your book is the key. Scan the code inside to unlock everything at curiouskidslearning.com — free, because you already own it: bonus printable activities, photos, videos, and more, plus a first look at what's coming next from Curious Kids, including Color and Learn books on the way. It's all waiting — come take advantage.

www.ingramcontent.com/pod-product-compliance
Lightning Source LLC
LaVergne TN
LVHW051131210926
848482LV00001B/67
* 9 7 9 8 8 9 9 9 8 0 0 0 8 *